BRITISH SHEEP BREEDS

Susannah Robin Parkin

SHIRE PUBLICATIONS
Bloomsbury Publishing Plc

Kemp House, Chawley Park, Oxford OX2 9PH, UK
29 Earlsfort Terrace, Dublin 2, Ireland
1385 Broadway, 5th Floor, New York, NY 10018, USA
Email: shire@bloomsbury.com

SHIRE is a trademark of Osprey Publishing Ltd

First published in Great Britain in 2015
Transferred to digital print on demand in 2023

A CIP record for this book is available from the
British Library.

Shire Library no. 803 – Print ISBN: 978 0 74781 448 1
ePDF: 978 1 78442 082 6 – Epub: 978 1 78442 081 9

Typeset in Garamond Pro and Gill Sans
Printed and bound in India by Replika Press
Private Ltd.

24 25 26 27 28   10 9 8 7

MIX
Paper from
responsible sources
FSC
www.fsc.org   FSC® C016779

The Woodland Trust
Shire Publications supports the Woodland Trust, the UK's
leading woodland conservation charity.

www.shirebooks.co.uk
To find out more about our authors and books visit our
website. Here you will find extracts, author interviews,
details of forthcoming events and the option to sign-up
for our newsletter.

COVER IMAGE
Cover design and photography by Peter Ashley. Front
cover: A Wiltshire Horn ram, Welland Colin, with
grateful thanks to Mike Adams. Back cover detail: A pair
of traditional hand shears. The Wiltshire Horn is a very
old native breed and, up until the end of the eighteenth
century, was the principal breed to be found on the
Wiltshire Downs. At the beginning of the nineteenth
century, George III distributed Merino sheep throughout
Wiltshire on the understanding that they would be crossed
with the Wiltshire Horn. However, this was met with
resistance from some local breeders and did not occur in
all the native flocks. Eventually, this wool-less breed lost
popularity in the latter part of the nineteenth century,
owing to the contemporary emphasis on wool production.

TITLE PAGE IMAGE
A Romney ewe and her lamb.

CONTENTS PAGE IMAGE
Jacob lambs.

IMAGE ACKNOWLEDGEMENTS
I would also like to thank the people who have allowed me
to use illustrations, which are acknowledged as follows:
Alamy, pages 11, 48, 51, 60 (top), 66, 67 (bottom); Jennifer
Batten, page 39; Kath Birkenshaw pages 9, 12–13, 16,
18, 44, 54, 57, 71; Kristina Boulden 58; Andew Boyce
(sheep's owner C. W. Morgan), page 50 (bottom); British
Wool Marketing Board pages 8; 34, 35, 45, 46 (top), 49,
50 (top), 56, 62, 64, 65, 67 (top); Richard Broad page
37; Kate Burrows, page 53; Sally Clarke, page 59; Dorset
Down Breed Society page 60; Simon Downham, page
44 (top); Getty Images, page 20; Michael Halliday, page
42; Hampshire Down Breed Society page 61; Douglas
Law, page 43; Richard Miller, page 69; North of England
Mule Association, page 15; Susannah Robin Parkin pages
25, 29, 46 (bottom); Rare Breeds Survival Trust page
38; Shutterstock, pages 3, 23, 30, 32 (both), 33, 40, 52,
57; Susan Schoenian pages 21, 22; Richard Small page
36; Wikicommons (Andreas Praefcke), Wikicommons
(Alexander Baxevanis), page 47; page 27 (top);
Wikicommons (Cstaffa), page 27 (lower).

ACKNOWLEDGEMENTS
The author would also like to thank members of the
National Sheep Association (NSA) for peer reviewing this
book. The NSA is a non-governmental organisation that
has been dedicated to representing the sheep industry since
1892. My gratitude extends to everyone that allowed me to
include their photographs, with particular thanks to Kath
Birkenshaw. Also, I would like to express my appreciation
to Katie Bates, Nicholas Parkin and Helen Alexandra
Parkin for all their advice, assistance and contribution
throughout the compilation of this book.

CONTENTS

THE HISTORY OF SHEEP AND SHEEP FARMING

BRITISH SHEEP THROUGH THE CENTURIES

ALTHOUGH IT APPEARS the dog was the first animal to be domesticated, it is thought that sheep followed the domestication process approximately two thousand years later. Domestication of sheep occurred around eleven thousand years ago and throughout history has played an integral role in British farming. Over the centuries sheep farming has faced many challenges, mainly owing to climate change and market demand. However, the evolution and adaptation of sheep breeds have ensured its survival.

Since prehistoric times sheep have provided mankind with fibre, pelts, meat and milk. This versatility combined with their behavioural characteristics is probably the major reason this species was so adaptable to domestication. Their origins can be traced to the wild Mouflon in Mesopotamia, a breed that is said to be the forebear of all sheep.

When the Romans first arrived in Britain in 55 BC, the sheep they found were a small primitive breed that had been introduced into Europe in the early Neolithic era (8000–6000 BC) – these subsequently became confined mainly to the island of Soay (pronounced 'so-ay'), from which the breed takes its name, and where the crofters traditionally made its fine wool into clothing. The Romans also brought their own long-woolled breed of sheep into Britain. Their sheep were small, hornless and white-faced, and their fleece was lustrous and wavy (crimp). It was these traits that the

Britons favoured, so farmers cross-bred the Roman sheep with the indigenous Soay and the development of the early native breeds began.

There is little information about the arrival of sheep similar to the Soay between about 4000 BC and the eighteenth century AD. However, it is acknowledged that traders, raiders, invaders and voyagers transported sheep, and thus the distribution of domesticated sheep began. These travellers introduced the northern European short-tailed breeds such as Boreray, Castlemilk Moorit, Manx Loaghtan, North Ronaldsay and Shetland. Travellers from the south of Europe, the Romans, contributed to the ancestors of the English longwools and the Down breeds such as Dorset Down, Hampshire Down, Oxford Down, Cambridge Down, Shropshire and Suffolk. These breeds were introduced in the first century AD, initially by trading and then through invasion.

After the departure of the Romans in AD 410, the Vikings invaded Britain between the eighth and tenth centuries, bringing with them their own sheep. The Viking breeds developed into the Blackface, Swaledale and Herdwick and it is from these three groups of primitive strains that the diverse variety of British breeds originated.

From the twelfth century until the seventeenth, Britain's textile industry flourished, sheep providing the wool for the cloth. During this period farmers had already started selective breeding to improve wool quality.

During the reign of Elizabeth I (1558–1603) the wool trade was the primary source of tax revenue and until the eighteenth century wool (rather than meat or milk) continued to provide the main source of income for the sheep farmer. It was the Industrial Revolution at the end of eighteenth century, accompanied by a population increase, which changed the course of sheep production. An era of high-production farming developed in which meat became

TEXTILES HAVE BEEN made by the tribes of northern Europe since 10,000 BC and, even before the arrival of the Romans and their sheep, Britain had a reputable wool industry. Great Britain became famous for textiles such as worsted, Scottish tweed and Axminster carpets. Many of the original British breeds had enormous influence, particularly in New Zealand, where the foundation flocks originated from Britain.

the main commodity, milk and wool becoming by-products. Selective breeding became more widely practised and characteristics different from those desired in Roman times prevailed.

During the Middle Ages the Spaniards bred their native sheep with sheep from the East, which were further improved by cross breeding with the Roman sheep that were introduced during their invasion. The result was a superior fleeced Merino which was worth much more than the courser British fleece.

The history of the Merino is fascinating. During the thirteenth and fourteenth centuries a powerful organisation known as the Spanish Mesta bred Merino sheep, which grazed on the southern plains in winter and the northern highlands in summer – a system called 'transhumance'. Merino wool, renowned to this day for its exceptionally high quality, was a very profitable commodity and the breeders gained many privileges from the kings of Castile. Furthermore, exportation of Merinos without royal permission was prohibited and was punishable by death, which safeguarded almost total dominance of the breed until the mid-eighteenth century. The British king, George III, an enthusiast farmer, was one of many sheep breeders who believed they could improve the wool still further by cross-

breeding the Merino with indigenous breeds such as the Wiltshire Horn. In 1789, he managed to smuggle some Merino sheep into Britain, which grazed the grounds at Kew. It became increasingly difficult for the Spanish Crown to uphold the law, and eventually the export ban was lifted – by the end of the eighteenth century Merino cross-breeding was at its height of popularity with British sheep breeders. Consequently, Merino bloodlines have been inherited by many native British breeds.

The agriculturist Robert Bakewell (1725–95), who farmed at Dishley in Leicestershire, became famous for his pedigree livestock breeding, in particular the new Leicester Longwool. The purpose of his selective breeding was to produce a cost-effective, early maturing breed. However, the livestock available to eighteenth-century sheep producers had already been selectively bred over several generations, albeit under less rigorous conditions. The desired characteristics

The Merino is the chief wool breed of sheep originating from Spain. They have white faces and legs. They can be recognised by their loose folds of skin and beautiful crimped fleeces. Rams have large curly horns.

during the earlier centuries were survival, local adaptation and ease of management rather than production. Nowadays the demand for meat-producing breeds has replaced the earlier desired traits.

Additionally, there was a reduction in demand for woollen textiles after the Industrial Revolution in the early nineteenth century. The reason for this decline was the introduction of other fibres such as cotton and, later, man-made fibres; these had a major impact on demand for woollen products.

By the twentieth century the demand for animal products dramatically increased, resulting in meat-producing breeds dominating the British market. European breeds such as Texel (from the Netherlands) have been imported to integrate into the British production system. These European breeds are cross-bred with native breeds to produce an early maturing, leaner and muscly lamb.

As a result, several of the original British breeds have become scarce. Nevertheless, many breeds have been saved

A Kerry Hill – once on the endangered list of the Rare Breeds Survival Trust, its numbers have recently increased to a safe level.

from extinction by the Rare Breeds Survival Trust. These rare sheep breeds can be found on smallholdings or conservational grazing on estates throughout Britain.

Among the breeds the Trust has safeguarded are the Welsh Mountain, Hebridean, Jacob, Kerry Hill, and more recently the Shropshire. Rare breeds were the ancestors of all British breeds and, although they are considered not to be economically viable in today's production system, it is important that rare-breed flocks are conserved to perpetuate traditional characteristics (genetic diversity). Equally important is the contribution they make to the conservation of the environment. While grazing, sheep create pathways, thus distributing seeds and insects. They are selective grazers, chosen for heath (such as lowland heathlands near York), the grasslands of southern England, woodland pastures, flood

The Texel originates from the island of Texel in the Netherlands and is a very popular breed in Britain. They were first imported in the 1970s and were cross-bred with native breeds such as the Lincoln, Leicester and Wensleydale.

plains and coastland marshes such as Romney, Kent. In this way they are suppressing areas of gorse, brambles or scrub, thus allowing more delicate species to survive.

In Great Britain today there are 32 million sheep representing ninety different breeds, all playing an important role in farming, from food production to conservation. This book introduces the reader to the most prominent among these.

SHEEP HISTORY AND ORIGINS

Domestic sheep originally evolved from seven wild species globally, but there are four breeds significant to British sheep: the Bighorn (North America), the Argali (central Asia), the Urial (south-western Asia) and the wild Mouflon (from western Asia, *Ovis orientalis*, and Europe, *Ovis musimon*). Other wild species include the Snow Sheep (from mountainous areas such as the Putoran Mountains in north-eastern Siberia) and the Thin Horn (British Columbia, Canada). However, it is the Mouflon that is particularly important to British sheep history as it is thought to be the ancestor of the Soay, a breed still found throughout Britain today. Relatively little is known of the evolutionary history of domestic sheep, their relationship to wild sheep species, or indeed how the first sheep arrived in Great Britain.

The ancient breeds of sheep originated from the desert or mountain grassland in the Fertile Crescent (western Iran and Turkey, Syria and Iraq) approximately eleven to twelve thousand years ago. Human settlements, livestock and arable farming developed simultaneously, probably owing to the lush pastures provided by these countries. Sheep were an easy species to domesticate because they grazed the vegetation within the vicinity of human settlements, making them easy to feed. The discovery of fossils in close proximity to human settlements in south-west Asia suggests that domestication of sheep may have first occurred there.

Although wild sheep existed in the northern and southern Levant, remains have rarely been encountered during the Epipalaeolithic era (*c.* 17,000–8500 BC). However, wild sheep bones have been recovered dating to the Proto-Neolithic era (*c.* 10,000–9000 BC).

Frontal view of skull with horns of *Ovis vignei*, a Urial wild sheep, from the Ladakh region of India.

The oldest domesticated sheep remains were discovered in northern Iraq (Zawi-Chemi-Chanidar), dating from around the Neolithic era (10,000 BC). Owing to these discoveries it is thought animals were first domesticated in the north-east of the Levant, around the Zagros Mountains, on the Turkish–Iranian border.

The different local environments, selective breeding and geographic location all contribute to regional differences and evolution of regional breeds. This, combined with selective breeding, led to the development of several breeds which, in

Previous pages:
Throughout the
north of England
the Swaledale
is an extremely
popular hill
breed. This breed
has proved to be
hardy and well
suited to the
harsh conditions
of their lofty
habitat.

the United Kingdom, have been classified into mountain, hill, upland and lowland breeds.

THE SHEEP INDUSTRY

In Great Britain the commercial sheep industry is based on a stratified three-tier system, and keepers select their stock to suit their particular altitude, grazing and production system.

The first tier is the hills and mountains of Scotland, Ireland, Wales and northern England, which require hardy mountain ewes, for example Scottish Blackface and Swaledale, as they can survive the harsh weather conditions and cope with poorer diets. Under these harsh conditions the ewes are bred for desirable characteristics such as mothering and milking abilities. Mountain and hill breeds tend to have one lamb per season and generally breed only for four seasons. The older ewes are transferred to the milder climate of lower areas, where they are crossed with longwool breeds. The improved environment gives a better quality of life to the older ewe, which in turn prolongs her life, enabling her to breed for a further two or three lamb crops. The cross results in a wide variation of half-bred sheep (known as Mules).

The female offspring of the mountain and hill breeds are retained as pure breeds for breeding stock. The male lambs and any surplus females are sold to upland or lowland farms to be finished or reared for meat production.

The second tier relates to the upland areas found in many parts of the United Kingdom, where specific breeds include Border Leicester and Bluefaced Leicester. The older draft mountain ewes are crossed with the longwool breeds to produce a wide variation of half-bred sheep or Mules (Bluefaced Leicester × Swaledale). The resulting ewe lambs are sold to the lowlands to be crossed with a lowland breed. Slower-growing lambs join the male 'store' lambs that have

The Mule is a medium-sized popular cross-bred sheep. This dam is renowned for her good mothering abilities, showing kindness and concern for her offspring. Other outstanding qualities are hardiness, thriftiness and longevity.

arrived from hill and upland areas to be finished (fattened ready for the butcher) on root crops such as swede over the autumn and winter months. The upland breeds are selected for their high fertility and milking ability.

The third tier relates to the lowland areas of southern England and other parts of the United Kingdom, where the Mules (cross-breeds) are bred with lowland rams to produce prime (finished) lambs for slaughter. Slower-growing lambs join the store lambs that have arrived from the hill and upland areas to be finished on root crops, such as swede or stubble turnips, over the autumn and winter months. The farmer's income on the lowland comes from the sale of prime finished lambs, breeding ewes and pedigree terminal sires.

This stratified system makes economic sense because it utilises the natural resources of British topography, which might otherwise be redundant; it also makes biological sense because it makes use of the hybrid vigour, whereby the progeny inherits the best qualities from both parents.

The Suffolk is a very popular lowland breed. They have a distinctive jet-black head and legs, with a dense white fleece. They evolved by crossing a Southdown ram with a Norfolk Horn ewe and were recognised as a pure breed in 1810.

CLOSED FLOCK SYSTEMS

While this unique stratified system is well tried and tested and forms the mainstay of British sheep production, there are also many sheep farmers who operate 'closed flocks'. The principle here is to implement a breeding policy that allows breeding stock to be retained alongside other livestock that are sold.

CROSS	SIRE BREED	DAM BREED
Northern Mule	Bluefaced Leicester	Swaledale
Scottish Mule	Bluefaced Leicester	Scottish Blackface
Welsh Mule	Bluefaced Leicester	Welsh Mountain
Clun Mule	Bluefaced Leicester	Clun
Greyface	Border Leicester	Scottish Blackface
Masham	Teeswater Wensleydale	Dalesbred, Roughfell Swaledale
Sottish Half-bred	Border Leicester	Cheviot
Welsh Half-bred	Border Leicester	Welsh Mountain

The most common cross-breeds and their derivations are illustrated below:

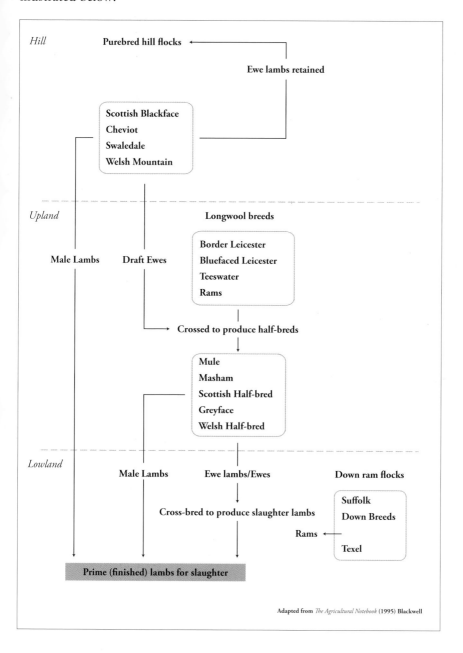

Hill   **Purebred hill flocks**

Ewe lambs retained

Scottish Blackface
Cheviot
Swaledale
Welsh Mountain

Upland   **Longwool breeds**

Border Leicester
Bluefaced Leicester
Teeswater
Rams

Male Lambs   **Draft Ewes**

Crossed to produce half-breds

Mule
Masham
Scottish Half-bred
Greyface
Welsh Half-bred

Lowland

Male Lambs   **Ewe lambs/Ewes**   **Down ram flocks**

Suffolk
Down Breeds

Cross-bred to produce slaughter lambs

Rams

Texel

Prime (finished) lambs for slaughter

Adapted from *The Agricultural Notebook* (1995) Blackwell

SHEEP BEHAVIOUR AND WOOL CHARACTERISTICS

BEHAVIOUR

EVER SINCE PEOPLE have kept sheep they have watched them and understood their biological and behavioural traits. The biological characteristics of sheep show that both ram and ewe will naturally come into season generally in September, when the days start to shorten. A few breeds come into season in summer (Dorset Horn, for example), and a few can breed around the year, such as Finnish Landrace. The ewe is pregnant for approximately 145 days and, depending upon location, normally gives birth to one to two lambs per season. The top commercial flocks will wean more than two lambs per ewe per season. At birth, lambs weigh approximately 4–6 kg and they are weaned from their mother at twelve weeks, when they are able to fend for themselves. At eighteen months old, the ewe is able to breed and will do so for approximately four to six years.

Over centuries, the shepherd's intuition and understanding of sheep behaviour has helped sheep to prosper under domesticated conditions. In recent years the study of animal behaviour has become a fascinating and popular science known as 'ethology'.

When the ewes come into oestrus they become restless. At this time the rams become aggressive and will often fight with each other. However, when the ram meets the ewe, both wild and domesticated sheep show affection by nuzzling.

The odour of the oestrous ewe stimulates the ram, although it is the ewe that seeks out the ram and stays close beside him.

Fleeces in the United Kingdom have incredibly varied characteristics. For this reason they are categorised into seven groups, depending on certain features:
1) fine wool
2) medium wool
3) coarse wool
4) Longwool
5) Mountain
6) Hill
7) naturally coloured wool.

Dall sheep (*Ovis dalli*). The senses play an important role in sheep reproduction. The reproductive cues are predominantly olfactory (smell). The ram is able to detect a ewe in oestrus by sniffing and the flehmen response is a reaction to several odours.

The male responds to the oestrous female by sniffing, extending the neck and curling the lip. This is called the 'flehmen response'. The tongue goes in and out and the male may bite the ewe's wool and will raise and lower one front leg in a stiff-legged striking motion. If the ewe is receptive she will stand for copulation (tupping).

This behaviour is an important part of the mating process and a ewe is more likely to reject a ram if mounted without prior courtship. Generally, mating behaviour is displayed during the day, with the most activity in the early morning and late afternoon.

A ewe's oestrus cycle is repeated every three weeks. Some farms mark their ewes by placing a coloured fluid on the ram's brisket (breast), which leaves a mark on the ewe's rump during copulation; the colour is changed every three weeks as a guide to when the ewes are likely to lamb. This helps the shepherd to plan the nutritional requirements of the ewe.

The mothering abilities of most sheep are exceptional, with the parental behaviour being the sole responsibility of the ewe. Sheep are known to be promiscuous and there is no indication of pair bonding.

Ewes show signs of lambing by isolating themselves away from the flock and then starting to show signs of straining. This increases in frequency, with their heads pointing skywards while they strain. The waters break and the feet of the lamb appear first, followed by the head and then the rest of the body. On occasions, if the lamb is poorly presented, help can be given by the shepherd.

Once the lamb is born the ewe licks it and will call it, which immediately stimulates the lamb's senses – movement, sound, touch and vision. This maternal stimulus is fundamental to the lamb's survival.

Then the lamb starts to stand; this varies in time between breeds, the sex of the lamb and the difficulty experienced over lambing. It is vital for the lamb to suckle as soon as possible to ensure it has a good intake of colostrum – rich in protein and immune factors – which is released only for the first few days and precedes the production of true milk.

After birth the ewe will lick the birth fluids from the lamb, thus drying and stimulating it. This process engages the ewe's taste and smell (gustation and olfactory) senses and she is able to recognise her lamb by the time it suckles.

A ewe and her
lamb snuggling
together –
a successful
bonding.

The behaviour of the ewe towards its lamb changes after lambing: during the first week of life the lamb is allowed to suckle at any time; thereafter, as the lamb matures, the mother gradually restricts the frequency and duration of feeding.

The lamb always walks at the flank of its mother and when at rest they will lie together. Initially the lamb grazes in close vicinity to its mother but it will gradually increase the distance over time. As the lambs gain strength and confidence they will congregate once or twice during the day and gambol (run or jump about playfully) but will eventually rejoin their mothers. At approximately six weeks old they learn to eat grass and eventually lambs will wean themselves.

Sheep are diurnal (active during the daytime) and adult sheep graze for nine to thirteen hours, followed by periods when they lie down and ruminate or rest. Their longest grazing spells occur in the morning and evening. During the winter sheep will naturally eat less, owing to seasonal food scarcity; as a survival mechanism they demonstrate an automatic decline in appetite.

Sheep have the ability to process thoughts and emotions, and they have a long-term memory. These processes are important because they remember where to locate food and the selection of the best herbage. Another survival mechanism is their social behaviour and flock organisation. Sheep are animals with a 'fight or flight' instinct and as a general rule have a strong impulse to flock (although some breeds have a tendency to disperse, such as Soay, which still exhibit more flight behaviour due to their wild characteristics). This provides them with the security of safety in numbers against potential predators. Within the flock they establish a social hierarchy and a leader will be recognised. This hierarchy may influence mating rights, and priority to access food, water and shelter.

Sheep panoramic vision covers 300 to 330 degrees and they have binocular vision of 25 to 50 degrees. Sight is a vital part of communication and when sheep are grazing they maintain visual contact with each other. Each sheep throws its head to check the position of any other sheep. This constant monitoring is probably what keeps sheep in a mob as they move along grazing. Sheep become extremely stressed if kept in isolation, so they always need a companion to reduce stress.

Sheep have good memories, thus enabling them to remember the location of specific food types. When sheep become ill they seek herbage with remedial properties.

Sheep need to be given medicines and vaccines, to have their feet checked, to have excess wool removed from around the tail area (crutching/dagging), to be dipped for pest control, to be inspected for general health, and to have their lambs sorted and weighed prior to sale and shearing. All these shepherding tasks can be stressful to the sheep; a sound understanding of sheep behaviour can therefore assist a shepherd in handling their flock successfully.

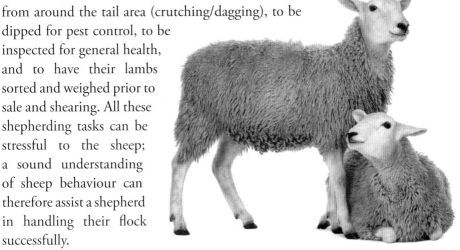

Many hours have been spent studying the behaviour of sheep to ensure that at all ages they can move quickly, easily and with the minimum of stress. It is much easier if the sheep move naturally and willingly through any handling system rather than being forced through. Several of the tasks are helped by the use of a trained sheepdog and the best can be very expensive to buy. Without their help man would never have domesticated sheep!

THE FLEECE

A sheep's fleece comprises three different fibres; these fibres are referred to as wool, kemp and hair.

Wool fibres vary in thickness, between 15 and 45 microns. The finest wools are to be found as an undercoat of the primitive breeds but man has bred sheep to suit the topography on which they are kept. Lowland sheep breeds tend to have finer, shorter wool (approximately 70 mm long), free from hair and kemp, whereas mountain breeds have longer, coarser wool (about 150 mm long) that contains hair and some kemp. Some Merino sheep have wool nearly as fine as cobwebs. Wool is virtually solid in nature, thus allowing dyes to be absorbed evenly. The thickness of the fibre varies slightly throughout the year and is influenced by feed and stress.

Kemp is the thickest of these fibres, having a large hollow central core or medulla. Kemp fibres are brittle and short and are normally considered to be commercially undesirable. They grow quickly and are constantly shed and regrown, so that they become caught up in the fleece but are not really part of it. They do not spin well because they are too brittle, slippery and small to be drawn into the thread, and because of their hollow core they do not take the dye as well as the wool fibres around them. Periodically these characteristics become fashionable, as has the interest in natural-coloured woollens and tweeds.

Hair, the third fibre in the fleece, grows among the fleece and invariably has a hollow core, depending on the condition of the sheep and the season of the year. Hair fibres are not commonly shed and tend to grow longer than wool fibres on the same sheep, so that they grow through the woolly coat. They then play the useful role of shedding the rain, especially in the hardy mountain breeds.

Besides the fineness of the fibre, the most important characteristic of wool is its crimp. This is the springiness in wool, caused by the waves and curls within the fibre. These can vary from being very tight waves in one plane, as in Merino wool, to waves incorporating a twist as in the longwool breeds. The twist may be greater than the wave component, resulting in spirals or ringlets, as in Wensleydale wool, and the fibres may be tightly waved but not held together. The finer the wool, the more crimps it has per centimetre.

A unique advantage of wool is that it is covered with overlapping scales, allowing the fibres to 'cling' together when spun. Under magnification, wool fibre looks similar to an elongated fir cone.

The wool fineness varies across the sheep but, as a rough guide, the finest wool is found at the neck and the coarsest at the britch (hind legs) and belly.

It is suggested that wool was the first fibre to be made into yarn in around 4000 BC. In AD 50 the Romans established the first wool factory, in Winchester. There are four steps involved when processing wool:
1) shearing
2) sorting and grading
3) spinning the yarn
4) manufacturing textile.

A white sheep is far easier to see on a hillside than a dark one, and white wool will absorb coloured dyes. So from the earliest days of domestication man selectively bred from those animals with light-coloured wool. The result is that almost none of our modern breeds are naturally coloured, with the exception of remnant populations of primitive breeds and individuals within white populations that happen to inherit a hidden recessive gene for colour from both parents and therefore appear coloured.

In ancestral wild sheep and the modern primitive breeds the fleece fibres stop growing for a couple of days in the early spring. The growth then resumes, but any period of non-growth (perhaps through stress, pregnancy or disease) can produce a break in the staple and a subsequent moult, or shedding of the coat. Man has overcome this by rearing healthy and appropriately nourished animals so that all the wool can be gathered at shearing (clipping) time. When the fleece has been harvested by shearers, it is rolled into a uniform bundle and then put into special bags (wool sheets or wool sacks), which are stored on the farm.

Shearing is a very skilled method of removing a sheep's fleece. There have been many different ways of handling a sheep to prevent it from struggling and make it comfortable. Originally the shearers used hand-operated blades to shear the sheep; now an electrically powered hand piece is used, which is much faster and more efficient. The method of handling sheep for shearing was perfected by the New Zealander Godfrey Bowen and is now practised by shearers around the world. The Bowen method of shearing consists of a series of complicated steps whereby the shearer uses the sheep's feet to rotate it clockwise. Simultaneously, the shearer uses his non-shearing hand to stretch the skin while the other uses the hand piece to cut the fleece.

The wool sheets or wool sacks are collected or delivered from the farm to one of the appointed British Wool

From about 1850 sheep shearing was performed with hand shears. These hand shears are still used in the twenty-first century for various tasks such as 'dagging' (taking the wool or dung from around the tail). As the wool trade boomed, faster methods of shearing were developed. The first shearing machine was invented by Lister in 1909.

Marketing Board (BWMB) sub-depots or main depots. At a sub-depot each farm's wool is weighed and bagged, then sent to a main depot.

At the BWMB depots the wool is graded by the immediate impression a fleece gives to an experienced professional wool grader as he assesses each fleece by look and feel. The grader looks for length of staple and softness, which is directly related to the fineness of fibres and the amount of crimp. He will look for levels of kemp and hair and check for fibre strength by giving it a pull. Any break or thinning in the growth of the wool during the year, due to ill health, bad weather conditions, an interruption of diet, mineral deficiency or bacterial attack, can cause staple tenderness or weakness. This is undesirable, because the wool might break in the spinning process, but it is soon detected in the simple test. The yield of useful wool from a fleece must also be considered. Wool is sold by weight, but up to half a fleece's weight will be waste products. Wool wax, dust, soil and moisture make up a large part, while the presence of paint, colouring and vegetation may make it necessary for sections to be downgraded or even discarded.

Some breeders dip their sheep in a colour (bloom dip) as they believe it enhances the overall appearance of the sheep at sales but this is not favoured by wool processors because the wool can then only be dyed a deeper colour and so is of less use to them.

After this grading the wools are placed with the similarly graded wool from other farms. Then each grade is pressed into bales ready to be auctioned.

The quality of British wool was commonly quoted in 'Bradford counts'. This was a method of measuring the fineness of sheep wool by how many 560-yard (512-metre) hanks of single strand yarn could be spun from 1 pound (454 grams) of top (a top is cleaned comb wool with the fibres placed parallel). So the finer the wool the more hanks would be produced; a Bradford count of 64 would produce 64 hanks of wool – more than 20 miles!

This is not a very good method to determine the fineness, as there are too many variables in the wool to make comparisons accurate, but it was the only way before the invention of modern, objective measuring systems. Now wools are measured in 'microns'; a count of 64 would be equivalent to 21 microns.

This old shearing machine is on display at the Seven Sisters Sheep Centre (East Sussex). Until the beginning of the twentieth century, sheep were arduously shorn with hand shears. However, in 1909 R. A. Lister & Co. started manufacturing shearing equipment in Dursley. This shearing machine, which was renowned for its ability to cut closer to the skin than hand blades, was just the beginning of a revolution which would forever change the way that sheep were shorn.

In Britain all wool is sold through the BWMB, which auctions the wool on behalf of the farmer. The BWMB is a farmer-run, non-profit-making cooperative established in 1950 to operate a central marketing system for all UK wool. The aim of the BWMB is to generate the best prices for its members. It is the only organisation in the world that collects, grades, sells and promotes wool. It is now the only remaining agricultural commodity board in the United Kingdom.

The farmer is paid a base price set by the BWMB upon delivery of their fleece. This fleece is then graded by BWMB and sold at auction on the farmer's behalf. The balance between the base price and the selling price, with handling expenses deducted, is forwarded to the farmer the following year.

In their heyday British sheep were exported worldwide and the country was renowned for its wool and textile industry. The wool made from British fleece is versatile and hard-wearing and is ideal for products such as clothing and rugs, but is predominantly used for carpets. In the twenty-first century, wool is still known for its unique properties, such as biodegradability.

ANCIENT AND PRIMITIVE BREEDS

ANCIENT BREEDS

ARCHAEOLOGICAL AND HISTORICAL evidence indicates that several wild species may have been domesticated or have contributed to modern domestic sheep breeds. Although the Bighorn has never been domesticated, scientific research suggests that the Mouflon contributed to the evolution of the European domesticated sheep and the Urial and Argali contributed to the evolution of Asiatic sheep. Knowledge of these wild ancestors helps to understand the behaviour and environmental adaptations of domestic sheep.

(EUROPEAN) MOUFLON (*OVIS MUSIMON*)

The rams of this breed have horns that curve forward while the ewes may be either horned or polled (no horns). The upper body has a dark reddish-brown fleece; darker in winter; dark median streak across the shoulders and adults have a greyish white saddle midway. The underneath of their body is paler.

ARGALI (*OVIS AMMON*)

The Argali has huge, majestic horns that can span more than a foot in circumference at the base and are from 3 to 4 feet in length, curving forward in a spiral with the tips of the horn facing outward. Both rams and ewes are horned.

The fleece of the upper body is fawn in summer but darkens in the winter; the lower body is pale. Adult rams have a very thick, large white throat ruff.

The European Mouflon (*Ovis musimon*).

The Argali
(*Ovis ammon*).

The Urial
(*Ovis vignei*).

URIAL (*OVIS VIGNEI*)

Both rams and ewes have horns that curve forward. The fleece of the upper body is grey in summer but dark brown in winter. There is a saddle patch and a large black throat ruff on adult rams; the lower body is pale/white.

BIGHORN (*OVIS CANADENSIS*)

Bighorn sheep have massive spiralled horns. The male horns are about 3 feet long and can weigh up to 14 kg. The size of these magnificent horns dictates the ram's position

Bighorn (*Ovis canadensis*).

It is believed that the Soay breed was introduced from central Europe by Neolithic farmers. It has remained commercially unimproved, with feral flocks living on the uninhabited islands of St Kilda.

within the hierarchy of the herd. Bighorn sheep are found on the grassy alpine meadows and foothills of the USA's mountain ranges. Their coat is hairy and the colouration differs according to the season. In the summer it is brown but it becomes paler in winter with an even paler patch on the rump.

PRIMITIVE BREEDS

SOAY

The Soay is the oldest of the primitive breeds and the only breed believed to have no connections with Viking introduction. The ancestors of the Soay are thought to

have been brought from central Europe by Neolithic farmers. On arrival in Britain, they became established in Soay, an island of the St Kilda archipelago in the Outer Hebrides, where they remain wild. In mainland Britain they are generally kept for ornamental purposes and conservation grazing.

They are a small sheep similar to the Mouflon, a very fine-boned and slow-growing breed. They always appear lean but are very fit. The tail is short and thin. The texture of their fleece can vary, from soft fine wool to coarser hairy fibre (kemp). They naturally moult their fleece in spring. Rams are two-horned, strong and handsome while the ewes are either two-horned or polled. Soay sheep are classified as 'vulnerable' on the RBST survival watch list.

HEBRIDEAN (ST KILDA)

The Hebridean (sometimes known as the St Kilda) and the Manx Loaghtan are both descended from the multi-horn

The Hebridean, a totally black breed (fading to brown in sunshine), has retained the magnificent multi-horn trait inherited from its Viking ancestors. This breed was largely superseded by the Scottish Blackface; consequently numbers dwindled, but populations are increasing again as flocks are nowadays kept to conserve upland areas.

This brown, multi-horned breed is native to the Isle of Man. During the twentieth century the Manx Loaghtan was threatened with extinction and it is thought to have been crossed with the Soay to restore population.

Viking sheep. They can be found in different colours: on the Hebridean islands they are black while on the Isle of Man only the moorit or tan colour survived. Both breeds have a short staple (cluster of wool fibres) and wool of medium fineness, and the fleece contains some hair and kemp. Both breeds have retained the magnificent Viking horns: some have two, others four and occasionally some have six. This hardy breed could once be found in the uplands of Scotland but the Scottish Blackface gained popularity and the Hebridean went out of fashion. However, the breed has survived these setbacks and is kept in conservation areas such as Spurn Point, East Yorkshire.

MANX LOAGHTAN

This sheep is another four-horned breed, native to the Isle of Man, and is one of the northern European short-tailed breeds. They can have up to four horns, and they have a brown face

and brown legs. The fleece is dark brown but may be paler on the outside and it may also vary in length, being either short or long.

There were fears of the breed becoming extinct during the twentieth century, so to ensure its existence the Manx Loaghtan may have been cross-bred with the Soay. Although the Manx Loaghtan is popular in Britain primarily for ornamental purposes, its grazing habits assist in maintaining the landscape and in conservation.

JACOB

The Jacob is the largest of all northern short-tailed breeds and can be either two-, four- or six-horned. The fleece is mottled in appearance, with the dark patches becoming lighter as the sheep matures. It is believed that it acquired its colour pattern from the Middle East and its horns from Viking stock in Spain before its introduction to Britain in the sixteenth century.

The Jacob is a large multi-horned breed with a fleece that is white with black or brown spots. It is not known how it arrived in the British Isles; but the lack of historical documentation suggests that it is a relatively new introduction.

Although the origin of the Jacob is unknown, it is thought that the first flocks in Great Britain were based on stock imported from the former Cape Colony.

The fleece of this breed is in demand for hand-made textiles as the range of colours produced is more varied than from other breeds.

SHETLAND

This is the smallest sheep of the British breeds, with a small, pretty tan face and legs. According to the Shetland Sheep Society – 1927 Breed Standard – both sexes may be either horned or hornless. The Shetland islanders developed their

This breed is native to the Shetland Isles and renowned for its fine, soft wool; the fleece comes in a variety of colours. Local people derived a thriving cottage industry from the wool until the Second World War; afterwards, meat production prevailed. Shetland remains a popular fleece for spinning and hand weaving, however.

small multi-coloured sheep to produce the finest wool of all indigenous British breeds.

The breed is found mainly on the Shetland Islands; however, it is believed to be of Scandinavian origin. The breed produces wool in several shades, including white, brown (moorit), grey and black. The wool is fine, soft and silky to the touch with a good, bulky down characteristic.

The Shetland wool industry survives (the wool is used for the famous 'Shetland' knitwear) and the breed is still farmed on the islands of Yell and Foula for this purpose.

NORTH RONALDSAY

Native to the Orkneys, this is a small breed that was maintained by crofters who specialised in growing crops. They built a wall around their island and put the sheep on the beach away from the crops. Here the sheep were forced to adapt to a diet of seaweed or kelp. The result is a unique, small, horned breed which produces meat and a small amount of wool. The fleece has a fine undercoat and a coarse hairy overcoat and comes in all colours.

North Ronaldsay is a small breed, indigenous to the Orkneys. The sheep survive on the seashore, living on a diet consisting entirely of seaweed.

MOUNTAIN, HILL AND UPLAND BREEDS

MOUNTAIN BREEDS

HISTORICALLY, THE MOUNTAIN sheep breeds of Britain fall into two principal groups and their combinations. The tan-faced breeds are derived from the medieval stock of marsh and hill sheep, while the black-faced mountain sheep were developed in the Pennines in the sixteenth and seventeenth centuries.

All the breeds in this group are very hardy and can withstand harsh weather conditions. They are intelligent sheep with a keen sense for survival. They carefully find shelter for themselves and their lambs and are aware of approaching storms. Female lambs that are to be kept in the flock are allowed to stay on the hill with their mothers. They live in a family group and learn the family's home range. Mountain sheep may travel many miles during a year but have a clear knowledge of their home and are said to be

MOST MOUNTAIN AND hill sheep are 'hefted'. This is a process whereby sheep are kept in an unfenced area of land. The shepherd feeds them on a particular area, to which the sheep will return. Over time this becomes a learned behaviour, passed from ewe to lamb over succeeding generations. Lambs graze with their mothers, who instil a lifelong knowledge of where optimal grazing and shelter can be found throughout the year.

Swaledale sheep can be found on the moorlands of the Pennines and are the mainstay of the country's sheep stratification system. They are active sheep, well adapted to the topography. Owing to the harsher conditions of their lofty home, ewes tend to give birth to only one lamb. This allows the ewe to provide sufficient milk for her offspring, giving the lamb a better chance of survival.

'hefted' on their hill. When a hill farm is sold, the hefted flock is sold with it. Should a new farmer try to buy ewes from his neighbours he would soon find them walking home to their old hillsides.

The male and surplus female lambs of mountain sheep are taken to richer farms to be fattened, or used in the stratified crossing system that enables British farmers to optimise the use of the country's landscapes.

BLACKFACE

The Blackface is the most abundant breed of sheep in Britain, making up 30 per cent of the total pure-bred sheep population. They are predominantly found in the mountainous regions of Scotland, Ireland and Northumberland. These tough animals are renowned for their ability to survive the harsh winters in Britain's most inhospitable areas, and, with their adaptability and versatility, they epitomise the mountain sheep. Anatomical differences have evolved to enable them to cope with their habitat, for example frame size, fleece length and texture. All Blackfaces have horns, with either black or black-and-white faces and legs.

Blackface is an extremely popular mountain breed. Its hybrid offspring are an important commercial sheep in lowland farming systems. The Mule retains the superior meat yield and high fecundity of the Bluefaced Leicester sire and the hardiness and good mothering ability of the Blackface ewe.

Records kept by monks show the breed originated in West Linton, in the Scottish Borders. It has survived for many centuries and in the nineteenth and twentieth centuries Britain saw the Blackface population rise, probably owing to breed improvement programmes. The Blackface has an important role in the sheep industry today, being positioned at the pinnacle of the stratification system. They provide a reservoir of progeny used in the cross-breeding process, producing prime lamb, and store lambs for finishing. In addition, their fleece may be used for carpets, mattresses or tweed, depending on the wool quality.

WHITE WELSH MOUNTAIN, BLACK WELSH MOUNTAIN

The Welsh Mountain is the second most common breed in Britain. The male has fine-looking spiralling horns but ewes are hornless. This is a small breed believed to have descended from the tan-faced sheep found in southern Britain throughout the Middle Ages. The South Wales Mountain can be found on the hills of Glamorgan, Monmouthshire, Carmarthenshire and south Powys and is bigger than its northern cousin. They have either a black or white fleece.

White Welsh Mountain.

Black Welsh Mountain lamb. Mountain and hill sheep are normally left with their naturally long tails to protect their udders from the cold wind, which can cause ailments such as mastitis.

The Black Welsh Mountain is the only purely black breed of sheep to be found in Britain. The fleece is fine wool but carries more dead hairs (kemps) than other breeds; however, both black and white fleeces are used for commercial textiles.

The Swaledale Sheep Breeders Society was founded in 1919 by a group of farmers living within a 7-mile radius of Tan Hill Inn, near where the counties of North Yorkshire, Durham and Cumbria meet. There are records that a ram sold in 2002 at the Swaledale three-day ram sale at Kirkby Stephen made £101,000 the highest price ever paid for a sheep in Britain.

SWALEDALE

The Swaledale has a black face with a white muzzle and is thought to be closely related to the Blackface. Like the Blackface, both rams and ewes have horns that are set low, and are round and wide. They are alert, active, and good

foragers, making the best use of the herbage provided by the fells and moorland.

One of the good qualities of the ewes is their ability to produce and rear offspring for three to four years under these lofty and harsh conditions. After this time they are drafted (transferred) to lowlands where conditions are more favourable. Once living in the milder climate and more nutritious pasture of the lowlands, they have the ability to produce offspring for a further three to four years.

Their coarse fleece varies in colour from white to grey with a dark patch at the nape of the neck. The wool is similar to the Blackface, which is extremely hard-wearing, and for this reason it is predominantly used in carpet, although some wool is occasionally used for hand-knitting.

LONK

Their face and lovely long legs are mottled black-and-white. Both rams and ewes have horns. The Lonk sheep breed is over two hundred years old and has been bred to cope with the hills of the central and south Pennines, in the north of England. The Lonk is raised to produce meat but also has a fleece that is suitable for carpets.

'Lonk' derives from the Lancashire word 'lanky', meaning long and thin – this is befitting as they are one of the largest native hill breeds in Britain.

Derbyshire
Gritstone.

The Herdwick
fleece is a
distinctive grey,
and is short,
thick and coarse.
Herdwick wool
is extremely
hard-wearing
and is used for
tweeds and
carpet making.

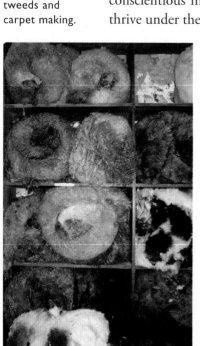

DERBYSHIRE GRITSTONE

This is a big, strong and handsome breed; they are alert and both sexes are hornless. Derbyshire Gritstone ewes are conscientious mothers and produce lively, hardy lambs that thrive under their doting mother's care.

The Derbyshire Gritstone sheep originated on the hills of the Dale of Goyt (now known as the Goyt Valley) on the edge of the Peak District around the year 1775 and was known in its early years as the 'Dale o'Goyt' sheep. The Derbyshire Gritstone sheep, which has evolved from the Dale o'Goyt sheep, is consequently one of the oldest British hill breeds.

Their dense, superior fleece is the finest of all the blackface type of sheep, suitable for the manufacture of high-class knitwear.

HERDWICK

The Herdwick is a small, fine-boned sheep with a white face and legs. The rams are horned but the ewes are not.

They originate from and still flourish in the Lake District, although small flocks can be found throughout the British Isles. While their origins cannot be confirmed, there are a couple of theories: one suggestion is that Scandinavian raiders brought them when they invaded Britain in the eighth century and the sheep remained after the Vikings left in the eleventh century. This would account for their name, thought to derive from thirteenth-century Norse meaning 'sheep pasture'. Another story suggests that a Spanish galleon capsized near the Cumbrian coast, and that the sheep on board swam ashore and survived. Once on dry land, the sheep found an environment they liked and settled there.

Herdwick sheep survive in rigorous conditions and craggy countryside. They manage to produce vivacious lambs with high survival traits.

HILL BREEDS

The hill breeds are pure strains of sheep that fit into roughly the same commercial niche as the half-bred daughters of mountain ewes by long-wool rams. They are hardy breeds and have been developed in the hilly regions, which they still traditionally occupy. They are quite prolific, often rearing healthy twins, and carry a good fleece. They are bred to rams of their own breed and their daughters are kept or sold as breeding sheep, while the male lambs are fattened for meat. Alternatively the ewes are mated to a down breed of ram to produce large crops of fast-growing good-quality butchers' lambs.

CHEVIOT

The Cheviot sheep get their name from their native Cheviot Hills, on the borders between northern England and Scotland. They can be identified by their distinctive white faces. Both the ram and ewe are hornless.

There is a strong possibility that Cheviot sheep have bloodlines stemming from Merino sheep that were probably shipped from the Continent into the port of Berwick in 1480 and again in 1560.

It is suggested that, as the church owned considerable portions of land in the hills, the care of the best class of sheep may have been with the monks attached to Kelso Abbey. Over centuries the breed has been selectively bred to create the Cheviot that is popular on the hills today. It is from this breeding by pioneers such as James Robson that the Cheviot is thought to stem, and it is recorded that in the eighteenth century Sir John Sinclair strongly recommended this strain of sheep for the north of Scotland, where their popularity grew.

With the development of cloth manufacturing and the increased importance of the wool-growing industry in the

Border towns of Selkirk, Galashiels and Hawick, so came the need for a good durable wool. So it was established that the Cheviot wool surpassed any other in the making of tweed and carpets. Although Cheviot wool is of great importance to the tweed industry, the primary purpose of this breed remains in the production of meat.

HILL RADNOR

This is an attractive sheep with grey nose, tan face and legs that are free from wool. The rams have elegant curved horns that spiral outwards but the ewes are hornless. Although gay in appearance, this is a hardy, attractive sheep and, like many hill breeds, they are good foragers. They can be found in the central marches of Wales, and the breed remains predominantly in the area around Brecon.

Their fleece is very popular with local hand-spinners and weavers owing to its fine staple. It is a lovely white fleece, having a dense structure.

The Hill Radnor may be used to cross-breed as the ewes are renowned for their strong maternal instinct. Additionally, when crossed with other breeds they can produce productive half-bred ewes or prime-quality lamb from either hill or lowland pastures.

The Clun Forest is a versatile breed that provides the market with a variety of products, being popular for its fleece, meat and for the provision of milk.

CLUN FOREST

The Clun Forest is a medium-sized, strikingly good-looking sheep, with dark legs and a narrow, dark face. They have 'pricked-up' ears, giving them an intelligent appearance, with a charming woolly top-knot on an otherwise hairless face. Neither the ram nor the ewe has horns. These sheep live in the loftier areas of Shropshire, on the Powys border.

Their fleece is beige in colour but fine, of superior quality and therefore popular with hand-spinners and weavers. In addition, owing to the high butterfat content in their milk, Cluns are often crossed with dairy sheep, such as the East Friesian, and provide dairy products.

A Kerry Hill, showing the breed's distinctive black-and-white colouring.

KERRY HILL

The Kerry Hill is a handsome sheep. Although hornless, its black nose and the sharply defined black-and-white mottled markings on its head and legs render it a very distinctive breed. Since the early nineteenth century they have lived on the English/Welsh borders, originating from a village called Kerry (near Newtown), from which their name is derived.

They have a good-quality fleece that is white, dense and free from kemp. The breed produces a pleasant meat, but flavour improves when the Kerry Hill is crossed with any of the hill, long wool or down breeds.

UPLAND BREEDS
BORDER LEICESTER

The Border Leicester is a descendant of the famous Dishley Leicester, bred by Robert Bakewell (1726–95), and, because of this, their history is well documented. In 1767 two brothers, George and Mathew Culley, established a flock of Border Leicesters in Northumberland and their stock originated from Bakewell's flock.

Over time two distinct types of Dishley Leicester evolved. Originally the brothers were crossing their sheep with the Teeswater, while other farmers on the northern borders were crossing their sheep with the Cheviot. As a result of this breeding variation, two distinct types evolved, which became

Flocks of Border Leicester are widely distributed throughout the British Isles. They may be bred 'pure' or crossed with any hill breed to produce a half-bred lamb. Their popularity extends beyond Great Britain: they are known to have been exported worldwide.

known as 'Bluecaps' and 'Redlegs'. The 'Redlegs' became the favoured type for the Border farmers, because of their hardier characteristics, and by 1850 this derivation became known as the Border Leicester.

The Border Leicester is still a very popular crossing sire and can now be found throughout Great Britain.

BLUEFACED LEICESTER

This breed has a magnificent, large Roman nose, alert eyes and long upright ears. The unusual but attractive face is dark blue with a covering of white hair.

The Bluefaced Leicester, or Hexham Leicester, as it is sometimes known, originated in Hexham, Northumberland, during the nineteenth century. By the 1930s the breed gained a reputation for being a good-quality crossing sire and consequently became a popular breed throughout the north of England. It has maintained popularity throughout the centuries and is still used to sire Blackface or Swaledale ewes.

The Bluefaced Leicester grows a fleece that is long, fine and dense, and is either white or natural brown in hue.

The Bluefaced Leicester ram is a sire of the Mule. One of the outstanding qualities of this breed is the ability to produce 'hybrid vigour' (inheriting the best traits from both parents) in the cross-bred offspring.

Their fleece tends to be quite small but provides a lustrous, resilient yarn.

TEESWATER

This hornless breed has either a white or greyish blue face with dark markings around the nose and ears and a characteristic top-knot falling over its face. The body looks long and they are tall but their most striking feature is their gorgeous fleece, which is long, fine and curly with a natural white lustre. For two hundred years they have been bred by farmers in Teesdale, County Durham, and have became renowned as the sire of the Masham cross-bred.

The Teeswater is renowned for its longevity and good milking ability, so ewes are able to provide sufficient milk to rear their lambs. Another excellent attribute is their high-quality fleece.

LOWLAND AND DOWN BREEDS

LOWLAND BREEDS

THE LOWLAND BREEDS are the third and final tier of the stratification system where sheep produce cross-bred offspring purely for meat production. The life of the lowland sheep is less severe than for those inhabiting higher regions of the country, the pasture being more luxuriant and nutritious. As a result they are very prolific, producing twins, triplets and often quadruplets, which the mother is generally able to rear owing to her abundance of milk.

Rams are pure breeds who are mated to cross-bred ewes. These ewes have a breeding lifespan of approximately seven years; the lambs are generally fast-growing and are sold for meat at approximately twelve weeks old.

Lowland sheep are able to withstand intensive grazing, making management easier for the shepherd.

LLEYN

The Lleyn (pronounced 'Cleen') ewe is a lovely white hornless sheep with a very feminine face. However, their appearance is deceptive: they are hardy, originally bred to thrive on the Lleyn peninusula, on the north-west coast of Wales. The Lleyn ewe is renowned for her mothering and good milking abilities. In their heyday, the Lleyn provided milk and cheese as well as rearing multiple lambs. As a result of breeding with Irish stock, which apparently 'improved' the breed, they continue to produce lambs for meat production. They grow a white,

The Suffolk was recognised as a pure breed in 1810. Today, the ram is used as a terminal sire to cross-breed with lowland ewes. The resultant offspring are quick to mature and are for meat production.

Lleyn sheep are renowned for their exceptional mothering ability. They are docile, produce a high milk yield and have an excellent-quality white fleece.

dense fleece of high quality with plenty of crimp, which can be used for textile manufacture.

TEXEL

The Texel is a white, hornless breed with a broad head, white face and a jet-black nose. It is medium-sized, with a long, rectangular body and well-pronounced muscles. The breed originates from the island of Texel, off the north-west coast of the Netherlands. They were first imported into the United Kingdom in the 1970s and have been selectively bred by farmers to improve their potential as a terminal sire. This breed has existed since Roman times, and with a contribution from British sheep (at some point it is thought to have been bred with a New Leicester and a Lincoln Longwool) has maintained its popularity throughout Europe.

The outstanding quality of the Texel is its muscling and docility. They are bred primarily for meat production; however, additional income can be sought from their dense, white fleece.

SUFFOLK

Like most of the lowland breeds, the Suffolk is polled (no horns), and it has a handsome jet-black face and black legs. The Suffolk is a result of cross-breeding between a Southdown ram and a Norfolk Horn and it was in 1810 that this cross became a breed in its own right. Over the centuries the Suffolk has flourished as a terminal sire and, although mostly found in the south of England (for example, in Suffolk, Kent and

The dense wool that is harvested from the Texel is of medium quality. It is used for a variety of products (including carpets and knitwear) and is popular for hand-spinning and other crafts.

The Suffolk has been around for two hundred years. Breed improvements are continuing today to ensure the breed survives the ever-changing demands of the market.

Devon), the breed has been exported worldwide. The ram is bred with lowland ewes to produce a good-quality, early maturing lamb. Their fleece is short and dense, inheriting qualities from the Norfolk Horn fleece, which was used in original East Anglian cloths.

ROMNEY MARSH

Like most domesticated breeds, this sheep is named after its local area. The Romney, or 'Kent', as it is sometimes called, is a large-framed hornless sheep with a heavy white fleece, renowned for its characteristic woolly top-knot.

It is a multi-purpose breed providing meat and wool as a source of revenue for the farmer.

The cream, semi-lustrous fleece of the Romney has always been an important feature, right back to the heyday of wool smuggling in the seventeenth century. This very versatile fleece is popular with the textile industry and also for the production of high-quality carpets such as Axminster.

The landscape on the coastal plain of Romney Marsh is flat and exposed. These marshes have a rich clay soil, which is capable of sustaining the production of this large, intensively grazed breed.

DOWN BREEDS

The group known as the down breeds are used as terminal sires in the stratified sheep system. They have been selected exclusively for meat production: the conformation of the carcass, the muscle shape and their consistency have been developed to produce an ideal joint for the butcher.

The down breeds are all white, fine-woolled sheep with a short-staple wool that weighs approximately 2–3 kilograms. The wool is ideally suited to sculpting, so that animals prepared for show are clipped out to produce a square-looking sheep with a flat back and rounded buttocks. Although the judges automatically correct for this in their own minds, it is impossible to compare a ram in full show turnout with one in the fields without being impressed by the stockman's efforts. Just as all the longwools can be traced back to Robert Bakewell and the Dishley Leicester, all the down breeds are traced to John Ellman of Glynde in East Sussex. In the late eighteenth century he began selectively breeding his native shortwool sheep of the Sussex hills, working towards a small, fast-growing lamb. The result was the Southdown.

The down breeds differ from other breeds in that they are able to breed throughout the year, delivering two crops of lambs in an eighteenth-month period. If the ram services the ewe in May their lambs are born in October; this 'out of season' lamb may command a higher price owing to the reduced market availability at this time of year.

SOUTHDOWN

The Southdown is a charming, docile sheep with a placid temperament. Their face and small ears are covered in wool, giving them a 'teddy bear' appearance. They have short woolly legs and are compact and muscly with a broad back. They are native to the Sussex Downs, where they thrive on the intensive grazing systems employed by the lowland sheep farmer.

Southdown.

During the period between the mid-nineteenth century and the First World War the Southdown was a very popular sheep, particularly with the American market. The Americans preferred a smaller, woollier-headed Southdown, and these characteristics evolved to meet their requirements. Although a popular breed with the Americans, it was in New Zealand that the Southdown had the biggest impact, as it was the foundation sire for the 'Canterbury lamb'. Nowadays, in Great Britain, the Southdown has reverted to its larger, more active origins and this is partially due to the re-introduction of blood from New Zealand.

Dorset Down.

DORSET DOWN

The Dorset Down is a hornless, medium-sized sheep, with stocky build and a dark brown head and legs. They are strong, active sheep, renowned for their docile demeanour. They are found in many parts of Britain and are named after the county from which they originate.

The Dorset is a breed resulting from crossing the Hampshire and Wiltshire ewes with pure Southdown rams. The male lambs are usually slaughtered for consumption and replacement ewes are kept for breeding, while the older ewes are culled from the flock.

The fleece of the Dorset Down is short, fine and high-quality; it is used for knitting and weaving rugs.

HAMPSHIRE DOWN

The Hampshire Down has a rich, dark brown face and ears, with a bright eye. The original Hampshire breed was a large, long-horned sheep; however, the breed has evolved by crossing the Southdown with the Berkshire Knot. The Hampshire Down can be found in Wiltshire, Hampshire, Berkshire and other parts of the south of England.

As with all the down breeds, Hampshires are an established terminal sire for commercial flocks at the final stage of the three-tier stratification system. The ewes are known for their longevity, good mothering abilities and their high-volume milk production ('milkiness').

Their fleece is white, of moderate length, and close with a fine texture; it is used for felts and blending with other wools.

Hampshire Down.

LONGWOOL AND CROSS-BREEDS

LONGWOOL BREEDS

In AD 43 the Romans brought the longwool breeds into Britain when they invaded the country. These breeds all produce large quantities of wool, for which they are famous, but they are also slow growing, producing very tasty meat. Longwool breeds were once a valuable asset and in their heyday their wool was very profitable to the farmer. While the upland and mountain breeds have remained pure over the centuries, most of the ancient longwools have either disappeared or been preserved by smallholders, enthusiasts and the Rare Breeds Survival Trust.

COTSWOLD

The Cotswold is a large polled longwool breed with an imposing appearance, having a cheeky forelock of wool on a white face with dark skin on the nose. They have long slender legs, and are completely covered in a lovely, long fleece resembling white dreadlocks. The breed, thought to be a descendant from the woolly Roman variety, can be found grazing the Cotswold Hills.

Originally, the Cotswold was specially bred for its long wool and at the height of the wool boom the fleece was a major export commodity. In their heyday, the income generated from the fleeces of Cotswold sheep was responsible for the famous 'wool churches' – wealthy wool merchants funded the building of many beautiful, ornate churches, for which the Cotswolds and East Anglia are renowned.

The Wensleydale (see page 65). Archaeological evidence shows the longwool breeds to be large-framed sheep. All the longwools fulfil similar roles of producing large quantities of wool as well as large mutton carcasses.

The renowned fleece from the Cotswold sheep is used today for specialist crafts, such as worsted spinning. Owing to the popularity of traditional crafts, their fleece has seen a surge of interest from craftspeople all over the world.

The nineteenth century saw a change in fashion and by the twentieth century the Cotswold's popularity declined in favour of the more muscular breeds.

LEICESTER LONGWOOL

The Leicester Longwool is a large, polled sheep with a bluish face, whose crown is covered with wool. Their fleece may be white and black and it is a fabulous, lustrous fleece popular with spinners, and used for hand-spinning, knitting and rugs.

This breed inhabits the Midland counties but has also travelled north into the Wolds of East Yorkshire and the hills of North Yorkshire. The Leicester Longwool (like all the Leicester breeds) is a descendant of the Dishley or New Leicester. They are, possibly, the most hybridised breed of all the British varieties and owe their existence to the agriculturist Robert Bakewell. Although the actual origins of Bakewell's Leicesters are unknown, the old Lincoln, Teeswater and Warwickshire are a few of the breeds named. In addition it is thought that crosses with the Ryeland, the South Down and other short-woolled breeds have been instrumental in the production of the Leicester Longwool.

Leicester Longwool – this breed is on the RBST endangered list. This is because of the fall in wool prices during the twentieth century and a movement away from slow-maturing mutton in favour of early maturing lean lamb.

WENSLEYDALE

The Wensleydale (see page 62) is a tall, white-woolled sheep with a head and ears that are blue. They sport a curly forelock that should be left intact at shearing. These sheep originate from North Yorkshire early in the nineteenth century and result from crossing a now-extinct local longwool breed from the region of the River Tees and the Dishley Leicester ram named 'Bluecap'. Bluecap was born in 1839 in the hamlet of East Appleton, near Bedale in North Yorkshire. His desirable features were his dark skin and superb quality of wool.

Wensleydale fleece is long and lustrous, each staple being curled throughout its length. This fleece is popular for spinners, is of good quality, and is therefore used for textile manufacture.

CROSS-BREEDS

A cross-breed is the progeny produced from two purebred sheep of different breeds; for example a North Country Mule is the offspring of a Bluefaced Leicester and a Swaledale. In Britain, Mule ewes are the backbone of the commercial sheep industry.

The North Country Mule is an elegant, majestic ewe with several outstanding qualities. Attributes include prolificacy, often giving birth to twins or triplets, good mothering skills, and a high milk yield that enables the Mule competently to rear her multiple lambs.

NORTH COUNTRY MULE

The North Country Mule is a pretty sheep with a black-and-white or brown-and-white mottled face and legs that match the face.

This cross-bred sheep is sired by a Bluefaced Leicester ram with a Swaledale or Northumberland-type Blackface ewe. The result of this cross is offspring renowned for their 'hybrid vigour' (retaining the best qualities of both breeds). The Mule is an elegant cross-breed renowned for mothering abilities, prolificacy, milkiness and hardiness. They are a popular commercial ewe found throughout Britain.

MASHAM

The Masham (Teeswater ram × Dalesbred/Blackface ewe) has been bred for over a century on the hills in the north of England. The Masham ewe is medium-sized and hornless and, like all cross-breeds, is bred to inherit the best characteristics from both parents. The desired traits of this breed are longevity, hardiness, milking ability, high prolificacy and strong mothering instincts.

Their fleece is very long and lustrous and designed to cope with the weather in the north of England.

Mashams are a result of a cross between a Teeswater sire and a Dalesbred, Swaledale or Scottish Blackface dam. They are mainly located in the north of England. Besides being good mothers, they produce a good-quality fleece.

SCOTCH MULE

The Scotch Mule has a proud black-and-white mottled face, ears that are 'pricked up' and alert and intelligent eyes. They are fine-boned and very hardy. The Scotch Mule (Bluefaced Leicester ram × Scottish Blackface ewe) is the backbone of the commercial sheep industry.

Their fleece is white with a medium staple, lustrous with a lovely crimp. It is good for spinning and is used for a variety of clothing products.

The majority of cross-bred ewes in Great Britain are sired by the Bluefaced Leicester. The Scotch Mule, like other cross-breeds, is highly prolific, thus creating financial benefits for the farmer.

FURTHER
INFORMATION

BOOKS

Alderson, Lawrence. *Rare Breeds*. Shire Publications, fourth edition, 2001.

British Wool Marketing Board. *British Sheep and Wool*. British Wool Marketing Board, third edition, 2010.

Clutton-Brock, Juliet. *Domesticated Animals*. London, British Museum (Natural History), first edition, 1981.

Jensen, P. *The Ethology of Domestic Animals*. CAB International, 2009.

Lynch, J. J., Hinch, G. N., and Adams, D. B. *The Behaviour of Sheep*. CAB International, first edition, 1992.

Mason, Ian. *A World Dictionary of Livestock Breeds, Types and Varieties*. CAB International, fourth edition, 1996.

National Sheep Association. *British Sheep*. NSA Publication, Worcester, ninth edition, 1998.

Ryder, M. L. and Stephenson, S. K. *Wool Growth*. Academic Press, 1968.

Upton, J., and Soden, D. *An Introduction to Keeping Sheep*. Farming Press, Ipswich, 1991.

Walling, Philip. *Counting Sheep*. Profile Books Ltd, London, 2014.

JOURNALS

The Sheep Farmer (journal of the National Sheep Association), The Sheep Centre, Malvern, Worcestershire WR13 6PH. Website www.nationalsheep.org.uk

The Journal for Weavers, Spinners and Dyers, 74 Huntingdon Road, Earlsdon, Coventry CV5 6PU. Website: www. thejournalforwsd.org.uk

WEBSITES

Most sheep breeds have their own breed society, all of which
can be accessed via the National Sheep Association. (Website:
www.nationalsheep.org.uk/sheep-breeds.php)
Department for Environment, Food and Rural Affairs
(DEFRA). Website: www.defra.gov.uk
English Beef and Lamb Executive (EBLEX).
Website: www.eblex.org.uk
National Animal Disease Information Service (NADIS).
Website: www.nadis.org.uk
Sustainable Control of Parasites in Sheep (SCOPS).
Website: www.scops.org.uk

ORGANISATIONS

British Wool Marketing Board, Wool House, Sidings Close,
Canal Road, Bradford, West Yorkshire BD2 1AZ.
National Sheep Association, The Sheep Centre, Malvern,
Worcestershire WR13 6PH.
Rare Breeds Survival Trust, Stoneleigh Park, Stoneleigh,
Warwickshire CV8 2LG.
Sheep Veterinary Society, Moredun Research Institute, Pentland
Science Park, Bush Loan, Penicuik, Midlothian EH26 0PZ.

PLACES TO VISIT

Acton Scott Working Farm Museum, Wenlock Lodge,
Acton Scott, Church Stretton, Shropshire SY6 6QN.
Telephone: 01694 781540.
Website: www.actonscott.com/index.php
Appleby Castle Conservation Centre (for rare breeds of
farm animals), Boroughgate, Appleby-in-Westmorland,
Cumbria CA16 6XH. Telephone: 01768 351402.
Cholmondeley Castle, Malpas, Cheshire SY14 8AH.
Telephone: 01829 720383.
Website: www.cholmondeleycastle.com
Cotswold Farm Park, Cotswold Farm Park Limited, Guiting
Power, near Cheltenham, Gloucestershire GL54 5UG.
Telephone: 01451 850307.
Website: www.cotswoldfarmpark.co.uk
Croxteth Hall and Country Park, Croxteth Hall Lane,
Liverpool L12 0HB. Telephone: 01512 333020.
Website: www.croxteth.co.uk
Easton Farm Park, Easton, Woodbridge, Suffolk IP13 0EQ.
Telephone: 01728 746475.
Website: www.eastonfarmpark.co.uk
Graves Park, City of Sheffield Recreation Department,
Meersbrook Park, Sheffield S8 9FL.
Telephone: 01142 500500.
Website: www.gravesparksheffield
Natural History Museum, Cromwell Road, London SW7 5BD.
Telephone: 020 7942 5000.
Website: www.nhm.ac.uk
Paradise Park, 16 Trelissick Road, Hayle, Cornwall TR27 4HB.
Telephone: 01992 470490.
Website: www.paradisepark.org.uk
Rare Breeds Centre, Highlands Farm, Woodchurch, Ashford,
Kent TN26 3RJ. Telephone: 01233 861493.
Website: www.rarebreeds.org.uk

Seven Sisters Sheep Centre, Gilberts Drive, East Dean,
   East Sussex BN20 0AA. Telephone: 01323 423207.
   Website: www.sheepcentre.co.uk
Totnes Rare Breeds Farm, Littlehempston, Totnes TQ9 6LZ.
   Telephone: 01803 840387.
   Website: www.totnesrarebreeds.co.uk
Shugborough Park Farm, The Shugborough Estate, Milford,
   Staffordshire ST17 0XB.
   Telephone: 0845 459 8900.
   Website: www.shugborough.org.uk/
   theshugboroughestate/Farm.aspx
The Sir George Staunton Country Park, Middle Park Way,
   Havant, Hampshire PO9 5HB.
   Telephone: 023 9245 3405.
   Website: www.visit-hampshire.co.uk/things-to-do/
   staunton-country-park-p46483

INDEX